Contents

L2A

L2A CONTENTS

Use of guidance

THE APPROVED DOCUMENTS

This document is one of a series that has been approved and issued by the Secretary of State for the purpose of providing practical guidance with respect to the technical requirements of the Building Regulations 2000 for England and Wales.

At the back of this document is a list of all the documents that have been approved and issued by the Secretary of State for this purpose.

Approved Documents are intended to provide guidance for some of the more common building situations. However, there may well be alternative ways of achieving compliance with the requirements. Thus there is no obligation to adopt any particular solution contained in an Approved Document if you prefer to meet the relevant requirement in some other way.

OTHER REQUIREMENTS

The guidance contained in an Approved Document relates only to the particular requirements of the Regulations that the document addresses. The building work will also have to comply with the requirements of any other relevant paragraphs in Schedule 1 to the Regulations.

There are Approved Documents that give guidance on each of the Parts of Schedule 1 and on Regulation 7.

LIMITATION ON REQUIREMENTS

In accordance with Regulation 8, the requirements in Parts A to D, F to K and N (except for paragraphs H2 and J6) of Schedule 1 to the Building Regulations) do not require anything to be done except for the purpose of securing reasonable standards of health and safety for persons in or about buildings (and any others who may be affected by buildings or matters connected with buildings). This is one of the categories of purpose for which Building Regulations may be made.

Paragraphs H2 and J6 are excluded from Regulation 8 because they deal directly with prevention of the contamination of water. Parts E and M (which deal, respectively, with resistance to the passage of sound, and access to and use of buildings) are excluded from Regulation 8 because they address the welfare and convenience of building users. Part L is excluded from Regulation 8 because it addresses the conservation of fuel and power. All these matters are amongst the purposes, other than health and safety that may be addressed by Building Regulations.

MATERIALS AND WORKMANSHIP

Any building work which is subject to the requirements imposed by Schedule 1 to the Building Regulations should, in accordance with Regulation 7, be carried out with proper materials and in a workmanlike manner.

You may show that you have complied with Regulation 7 in a number of ways. These include the appropriate use of a product bearing CE marking in accordance with the Construction Products Directive (89/106/EEC)[1], the Low Voltage Directive (73/23/EEC and amendment 93/68/EEC)[2] and the EMC Directive (89/336/ EEC)[3] as amended by the CE Marking Directive (93/68/EEC)[4] or a product complying with an appropriate technical specification (as defined in those Directives), a British Standard, or an alternative national technical specification of any state which is a contracting party to the European Economic Area which, in use, is equivalent, or a product covered by a national or European certificate issued by a European Technical Approval Issuing body, and the conditions of use are in accordance with the terms of the certificate. You will find further guidance in the Approved Document supporting Regulation 7 on materials and workmanship.

INDEPENDENT CERTIFICATION SCHEMES

There are many UK product certification schemes. Such schemes certify compliance with the requirements of a recognised document that is appropriate to the purpose for which the material is to be used. Materials which are not so certified may still conform to a relevant standard.

Many certification bodies that approve such schemes are accredited by UKAS.

TECHNICAL SPECIFICATIONS

Building Regulations are made for specific purposes: health and safety, energy conservation and the welfare and convenience of disabled people. Standards and technical approvals are relevant guidance to the extent that they relate to these considerations. However, they may also address other aspects of performance such as serviceability, or aspects which although they relate to health and safety are not covered by the Regulations.

---

[1] As implemented by the Construction Products Regulations 1991 (SI 1991/1620).
[2] As implemented by the Electrical Equipment (Safety) Regulations 1994 (SI 1994/3260).
[3] As implemented by the Electromagnetic Compatibility Regulations 1992 (SI 1992/2372).
[4] As implemented by the Construction Products (Amendment) Regulations 1994 (SI 1994/3051) and the Electromagnetic Compatibility (Amendment) Regulations 1994 (SI 1994/3080).

When an Approved Document makes reference to a named standard, the relevant version of the standard is the one listed at the end of the publication. However, if this version has been revised or updated by the issuing standards body, the new version may be used as a source of guidance provided it continues to address the relevant requirements of the Regulations.

The appropriate use of a product that complies with a European Technical Approval as defined in the Construction Products Directive will meet the relevant requirements.

The Office intends to issue periodic amendments to its Approved Documents to reflect emerging harmonised European Standards. Where a national standard is to be replaced by a European harmonised standard, there will be a co-existence period during which either standard may be referred to. At the end of the coexistence period the national standard will be withdrawn

THE WORKPLACE (HEALTH, SAFETY AND WELFARE) REGULATIONS 1992

The Workplace (Health, Safety and Welfare) Regulations 1992 as amended by The Health and Safety (Miscellaneous Amendments) Regulations 2002 (SI 2002/2174) contain some requirements which affect building design. The main requirements are now covered by the Building Regulations, but for further information see: *Workplace health, safety and welfare: Workplace (Health, Safety and Welfare) Regulations 1992, Approved Code of Practice, L24, HMSO, 1992* (ISBN 0 71760 413 6).

The Workplace (Health, Safety and Welfare) Regulations 1992 apply to the common parts of flats and similar buildings if people such as cleaners and caretakers are employed to work in these common parts. Where the requirements of the Building Regulations that are covered by this Part do not apply to dwellings, the provisions may still be required in the situations described above in order to satisfy the Workplace Regulations.

MIXED USE DEVELOPMENT

In mixed use developments part of a building may be used as a dwelling while another part has a non-domestic use. In such cases, if the requirements of this Part of the Regulations for dwellings and non-domestic use differ, the requirements for non-domestic use should apply in any shared parts of the building.

The Requirement

This Approved Document, which takes effect on 6 April 2006, deals with the energy efficiency requirements in the Building Regulations 2000 (as amended by SI 2001/3335 and SI 2006/652). The energy efficiency requirements are conveyed in Part L of Schedule 1 to the Regulations and regulations 4A, 17C and 17D as described below.

Requirement	Limits on application

Part L Conservation of fuel and power
 	
L1. Reasonable provision shall be made for the conservation of fuel and power in buildings by:
 	
    a. limiting heat gains and losses:
 	
        i. through thermal elements and other parts of the building fabric; and
 	
        ii. from pipes, ducts and vessels used for space heating, space cooling and hot water services;
 	
    b. providing and commissioning energy efficient fixed building services with effective controls; and
 	
    c. providing to the owner sufficient information about the building, the fixed building services and their maintenance requirements so that the building can be operated in such a manner as to use no more fuel and power than is reasonable in the circumstances.	

Other changes to the Regulations

There are new Regulations that introduce new energy efficiency requirements and other relevant changes to the existing regulations. For ease of reference the principal elements of the regulations that bear on energy efficiency are repeated below and, where relevant, in the body of the guidance in the rest of this Approved Document. However it must be recognised that the Statutory Instrument takes precedence if there is any doubt over interpretation.

Interpretation

Regulation 2(1) is amended to include the following new definitions.

'Change to a building's energy status' means any change which results in a building becoming a building to which the energy efficiency requirements of these Regulations apply, where previously it was not.

'Energy efficiency requirements' means the requirements of regulations 4A, 17C and 17D and Part L of Schedule 1.

'Fixed building services' means any part of, or any controls associated with:

a. fixed internal or external lighting systems but does not include emergency escape lighting or specialist process lighting; or

b. fixed systems for heating, hot water service, air conditioning or mechanical ventilation.

'Renovation' in relation to a thermal element means the provision of a new layer in the thermal element or the replacement of an existing layer, but excludes decorative finishes, and 'renovate' shall be construed accordingly.

New paragraphs (2A) and (2B) are added to Regulation 2 as follows.

(2A) 'Thermal element' means a wall, floor or roof (but does not include windows, doors, roof windows or roof-lights) which separates a thermally conditioned part of the building ('the conditioned space') from:

a. the external environment (including the ground); or

b. in the case of floors and walls, another part of the building which is:

    i. unconditioned;

    ii an extension falling within class VII of Schedule 2; or

    iii. where this paragraph applies, conditioned to a different temperature,

and includes all parts of the element between the surface bounding the conditioned space and the external environment or other part of the building as the case may be.

(2B) Paragraph (2A)(b)(iii) only applies to a building which is not a dwelling, where the other part of the building is used for a purpose which is not similar or identical to the purpose for which the conditioned space is used.

Meaning of building work

Regulation 3 is amended as follows.

3.–(1) In these Regulations 'building work' means:

a. the erection or extension of a building;

b. the provision or extension of a controlled service or fitting in or in connection with a building;

c. the material alteration of a building, or a controlled service or fitting, as mentioned in paragraph (2);

d. work required by regulation 6 (requirements relating to material change of use);

e. the insertion of insulating material into the cavity wall of a building;

f. work involving the underpinning of a building;

g. work required by regulation 4A (requirements relating to thermal elements);

h. work required by regulation 4B (requirements relating to a change of energy status);

i. work required by regulation 17D (consequential improvements to energy performance).

(2) An alteration is material for the purposes of these Regulations if the work, or any part of it, would at any stage result:

a. in a building or controlled service or fitting not complying with a relevant requirement where previously it did; or

b. in a building or controlled service or fitting which before the work commenced did not comply with a relevant requirement, being more unsatisfactory in relation to such a requirement.

(3) In paragraph (2) 'relevant requirement' means any of the following applicable requirements of Schedule 1, namely:

    Part A (structure)
    paragraph B1 (means of warning and escape)
    paragraph B3 (internal fire spread – structure)
    paragraph B4 (external fire spread)
    paragraph B5 (access and facilities for the fire service)
    Part M (access to and use of buildings).

Requirements relating to building work

Regulation 4 is amended as follows

4.–(1) Subject to paragraph 1A building work shall be carried out so that:

a. it complies with the applicable requirements contained in Schedule 1; and

b. in complying with any such requirement there is no failure to comply with any other such requirement.

(1A) Where:

a. building work is of a kind described in regulation 3(1)(g), (h) or (i); and

b. the carrying out of that work does not constitute a material alteration,

that work need only comply with the applicable requirements of Part L of Schedule 1.

(2) Building work shall be carried out so that, after it has been completed:

a. any building which is extended or to which a material alteration is made; or

b. any building in, or in connection with, which a controlled service or fitting is provided, extended or materially altered; or

c. any controlled service or fitting,

complies with the applicable requirements of Schedule 1 or, where it did not comply with any such requirement, is no more unsatisfactory in relation to that requirement than before the work was carried out.

Requirements relating to thermal elements

A new regulation 4A is added as follows:

4A.–(1) Where a person intends to renovate a thermal element, such work shall be carried out as is necessary to ensure that the whole thermal element complies with the requirements of paragraph L1(a)(i) of Schedule 1.

(2) Where a thermal element is replaced, the new thermal element shall comply with the requirements of paragraph L1(a)(i) of Schedule 1.

Requirements relating to a change to energy status

A new regulation 4B is added as follows:

4B.–(1) Where there is a change to a building's energy status, such work, if any, shall be carried out as is necessary to ensure that the building complies with the applicable requirements of Part L of Schedule 1.

(2) In this regulation 'building' means the building as a whole or parts of it that have been designed or altered to be used separately.

Requirements relating to a material change of use

Regulation 6 is updated to take account of the changes to Part L.

Exempt buildings and work

Regulation 9 is substantially altered as follows.

9.–(1) Subject to paragraphs (2) and (3) these Regulations do not apply to:

a. the erection of any building or extension of a kind described in Schedule 2; or

b. the carrying out of any work to or in connection with such a building or extension, if after the carrying out of that work it is still a building or extension of a kind described in that Schedule.

(2) The requirements of Part P of Schedule 1 apply to:

a. any greenhouse;

b. any small detached building falling within class VI in Schedule 2; and

c. any extension of a building falling within class VII in Schedule 2,

which in any case receives its electricity from a source shared with or located inside a dwelling.

(3) The energy efficiency requirements of these Regulations apply to:

a. the erection of any building of a kind falling within this paragraph;

b. the extension of any such building, other than an extension falling within class VII in Schedule 2; and

c. the carrying out of any work to or in connection with any such building or extension.

(4) A building falls within paragraph (3) if it:

a. is a roofed construction having walls;

b. uses energy to condition the indoor climate; and

c. does not fall within the categories listed in paragraph (5).

(5) The categories referred to in paragraph (4)(c) are:

a. buildings which are:

i. listed in accordance with section 1 of the Planning (Listed Buildings and Conservation Areas) Act 1990;

ii. in a conservation area designated in accordance with section 69 of that Act; or

iii. included in the schedule of monuments maintained under section 1 of the Ancient Monuments and Archaeological Areas Act 1979,

where compliance with the energy efficiency requirements would unacceptably alter their character or appearance;

b. buildings which are used primarily or solely as places of worship;

c. temporary buildings with a planned time of use of two years or less, industrial sites, workshops and non-residential agricultural buildings with low energy demand;

d. stand-alone buildings other than dwellings with a total useful floor area of less than 50m².

(6) In this regulation:

a. 'building' means the building as a whole or parts of it that have been designed or altered to be used separately; and

b. the following terms have the same meaning as in European Parliament and Council Directive 2002/91/EC on the energy performance of buildings:

   i. 'industrial sites';

   ii. 'low energy demand';

   iii. 'non-residential agricultural buildings';

   iv. 'places of worship';

   v. 'stand-alone';

   vi. 'total useful floor area';

   vii. 'workshops'.

Giving of a building notice or deposit of plans

Regulation 12 is substantially amended as follows.

12.–(1) In this regulation 'relevant use' means a use as a workplace of a kind to which Part II of the Fire Precautions (Workplace) Regulations 1997 applies or a use designated under section 1 of the Fire Precautions Act 1971.

(2) This regulation applies to a person who intends to:

a. carry out building work;

b. replace or renovate a thermal element in a building to which the energy efficiency requirements apply;

c. make a change to a building's energy status; or

d. make a material change of use.

(2A) Subject to the following provisions of this regulation, a person to whom this regulation applies shall:

a. give to the local authority a building notice in accordance with regulation 13; or

b. deposit full plans with the local authority in accordance with regulation 14.

(3) A person shall deposit full plans where he intends to carry out building work in relation to a building to which the Regulatory Reform (Fire Safety) Order 2005 applies, or will apply after the completion of the building work.

(4) A person shall deposit full plans where he intends to carry out work which includes the erection of a building fronting on to a private street.

(4A) A person shall deposit full plans where he intends to carry out building work in relation to which paragraph H4 of Schedule 1 imposes a requirement.

(5) A person who intends to carry out building work is not required to give a building notice or deposit full plans where the work consists only of work:

a. described in column 1 of the Table in Schedule 2A if the work is to be carried out by a person described in the corresponding entry in column 2 of that Table, and paragraphs 1 and 2 of that schedule have effect for the purposes of the descriptions in the Table; or

b. described in Schedule 2B.

(6) Where regulation 20 of the Building (Approved Inspectors etc.) Regulations 2000 (local authority powers in relation to partly completed work) applies, the owner shall comply with the requirements of that regulation instead of with this regulation.

(7) Where:

a. a person proposes to carry out work which consists of emergency repairs;

b. it is not practicable to comply with paragraph (2A) before commencing the work; and

c. paragraph (5) does not apply,

he shall give a building notice to the local authority as soon as reasonably practicable after commencement of the work.

Regulations 13 (particulars and plans where a building notice is given) and 14 (full plans)

These are amended to apply to renovation or replacement of a thermal element and a change to a building's energy status.

New Part VA to the Regulations

Energy Performance of buildings

New Regulations are added as follows.

Methodology of calculation of the energy performance of buildings

17A. The Secretary of State shall approve a methodology of calculation of the energy performance of buildings.

Minimum energy performance requirements for buildings

17B. The Secretary of State shall approve minimum energy performance requirements for new buildings, in the form of target CO_2 emission rates, which shall be based upon the methodology approved pursuant to regulation 17A.

New buildings

17C. Where a building is erected, it shall not exceed the target CO_2 emission rate for the building that has been approved pursuant to regulation 17B.

Consequential improvements to energy performance

17D.–(1) Paragraph (2) applies to an existing building with a total useful floor area over $1000m^2$ where the actual building work consists of or includes:

a. an extension;

b. the initial provision of any fixed building services; or

c. an increase to the installed capacity of any fixed building services.

(2) Subject to paragraph (3), where this regulation applies, such work, if any, shall be carried out as is necessary to ensure that the building complies with the requirements of Part L of Schedule 1.

(3) Nothing in paragraph (2) requires work to be carried out if it is not technically, functionally and economically feasible.

Interpretation

17E. In this Part 'building' means the building as a whole or parts of it that have been designed or altered to be used separately.

Part VI – Miscellaneous

New Regulations are added as follows.

Pressure testing

20B.–(1) This regulation applies to the erection of a building in relation to which paragraph L1(a)(i) of Schedule 1 imposes a requirement.

(2) Where this regulation applies, the person carrying out the work shall, for the purpose of ensuring compliance with regulation 17C and paragraph L1(a)(i) of Schedule 1:

a. ensure that:

   i. pressure testing is carried out in such circumstances as are approved by the Secretary of State; and

   ii. the testing is carried out in accordance with a procedure approved by the Secretary of State; and

b. subject to paragraph (5), give notice of the results of the testing to the local authority.

(3) The notice referred to in paragraph (2)(b) shall:

a. record the results and the data upon which they are based in a manner approved by the Secretary of State; and

b. be given to the local authority not later than seven days after the final test is carried out.

(4) A local authority is authorised to accept, as evidence that the requirements of paragraph (2)(a)(ii) have been satisfied, a certificate to that effect by a person who is registered by the British Institute of Non-destructive Testing in respect of pressure testing for the air tightness of buildings.

(5) Where such a certificate contains the information required by paragraph (3)(a), paragraph (2)(b) does not apply.

Commissioning

20C.–(1) This regulation applies to building work in relation to which paragraph L1(b) of Schedule 1 imposes a requirement, but does not apply where the work consists only of work described in Schedule 2B.

(2) Where this regulation applies the person carrying out the work shall, for the purpose of ensuring compliance with paragraph L1(b) of Schedule 1, give to the local authority a notice confirming that the fixed building services have been commissioned in accordance with a procedure approved by the Secretary of State.

(3) The notice shall be given to the local authority:

a. not later than the date on which the notice required by regulation 15(4) is required to be given; or

b. where that regulation does not apply, not more than 30 days after completion of the work.

CO_2 emission rate calculations

20D.–(1) Subject to paragraph (4), where regulation 17C applies the person carrying out the work shall provide to the local authority a notice which specifies:

a. the target CO_2 emission rate for the building; and

b. the calculated CO_2 emission rate for the building as constructed.

(2) The notice shall be given to the local authority not later than the date on which the notice required by regulation 20B is required to be given.

(3) A local authority is authorised to accept, as evidence that the requirements of regulation 17C would be satisfied if the building were constructed in accordance with an accompanying list of specifications, a certificate to that effect by a person who is registered by:

a. FAERO Limited; or

b. BRE Certification Limited,

in respect of the calculation of CO_2 emission rates of buildings.

(4) Where such a certificate is given to the local authority:

a. paragraph (1) does not apply; and

b. the person carrying out the work shall provide to the local authority not later than the date on which the notice required by regulation 20B is required to be given a notice which:

   i. states whether the building has been constructed in accordance with the list of specifications which accompanied the certificate; and

   ii. if it has not, lists any changes to the specifications to which the building has been constructed.

Section 0: General guidance

CONVENTIONS USED IN THIS DOCUMENT

1 In this document the following conventions have been adopted to assist understanding and interpretation:

a. Texts shown against a green background are extracts from the Building Regulations as amended and convey the legal requirements that bear on compliance with the **energy efficiency requirements**. It should be remembered however that building works must comply with all the other relevant provisions. Similar provisions are conveyed by the Building (Approved Inspectors) Regulations as amended.

b. Key terms are printed in **bold italic text** and defined for the purposes of this Approved Document in Section 5 of this document.

c. References given as footnotes and repeated as end notes are given as ways of meeting the requirements or as sources of more general information as indicated in the particular case. The Approved Document will be amended from time to time to include new references and to refer to revised editions where this aids compliance.

d. Additional *commentary in italic text* appears after some numbered paragraphs. The commentary is intended to assist understanding of the immediately preceding paragraph or sub-paragraph, but is not part of the approved guidance.

TYPES OF WORK COVERED BY THIS APPROVED DOCUMENT

2 This Approved Document is intended to give guidance in relation to works comprising:

a. The construction of new buildings other than **dwellings**;

b. **fit-out works** where this is included as part of the construction of the building. (ADL2B is intended to apply to **fit-out works** in other circumstances.)

c. The construction of extensions to existing buildings that are not dwellings where the **total useful floor area** of the extension is greater than 100m² and greater than 25% of the **total useful floor area** of the existing building.

3 When constructing a building that contains **dwellings**, account should also be taken of the guidance in Approved Document L1A. In most instances, Approved Document L1A should be used for guidance relating to the work on the individual **dwellings**, with this Approved Document L2A giving guidance relating to the parts of the building that are not a **dwelling** such as heated common areas and, in the case of mixed-use developments, the commercial or retail space.

*It should be noted that **dwellings** refer to self-contained units. **Rooms for residential purposes** are not **dwellings**, and so Approved Document L2A applies to, for instance, boarding houses, hostels and student accommodation blocks.*

4 If a building that is to be used for industrial or commercial purposes (e.g. a workshop or an office) also contains living accommodation, it should be treated as a **dwelling** if the industrial or commercial part could revert to domestic use on a change of ownership. This could be the case if:

a. there is direct access between the industrial or commercial space and the living accommodation; and

b. both are contained within the same thermal envelope; and

c. the living accommodation occupies a substantial proportion of the total area of the building.

*Sub paragraph c) means that a small manager's flat in a large non-domestic building would not mean the whole building should be treated as a **dwelling**.*

TECHNICAL RISK

5 Building work must satisfy all the technical requirements set out in Regulations 4A, 4B, 17C, 17D, and Schedule 1 of the Building Regulations. Part B (Fire safety), Part E (Resistance to the passage of sound), Part F (Ventilation), Part C (Site preparation and resistance to moisture), Part J (Combustion appliances and fuel storage systems) and Part P (Electrical safety) are particularly relevant when considering the incorporation of energy efficiency measures.

6 The inclusion of any particular energy efficiency measure should not involve excessive technical risk. BR 262[5] provides general guidance on avoiding risks in the application of thermal insulation.

DEMONSTRATING COMPLIANCE

7 In the Secretary of State's view, compliance with Part L and regulation 17C would be demonstrated by meeting the five separate criteria as set out in the following paragraphs. Appendix A contains a checklist that can be used to confirm that all the criteria have been met satisfactorily.

---

[5]   BR 262 *Thermal Insulation: Avoiding Risks*, BRE, 2001.

The checklist can benefit both developers and building control.

8 Criterion 1: the calculated CO_2 emission rate for the building as constructed (the building emission rate, **BER**) must not be greater than the target rate (the target emission rate, **TER**) which is determined by following the procedures set out in paragraphs 18 to 23; and

This is required by Regulation 17C – see page 9.

9 Criterion 2: the performance of the building fabric and the heating, hot water and fixed lighting systems should be no worse than the design limits set out in paragraphs 33 to 62; and

10 Criterion 3: Those parts of the building that are not provided with comfort cooling systems have appropriate passive control measures to limit solar gains. The guidance given in paragraphs 63 to 65 of this Approved Document provide a way of demonstrating that suitable provisions have been made; and

*The aim is to counter excessive internal temperature rise in summer in accommodation that does not need air conditioning and where it is not therefore provided. The impact on CO_2 emissions from mechanically cooled buildings is taken into account in the **BER**.*

Paragraphs 9 and 10 address the requirements in Part L1(a) and L1(b).

11 Criterion 4: the performance of the building, as built, is consistent with the prediction made in the **BER**. The procedures described in Section 2 can be used to show this criterion has been met; and

12 Criterion 5: The necessary provisions for enabling energy efficient operation of the building are put in place. The procedures described in Section 3 can be used to show this criterion has been met.

Paragraph 12 addresses the requirement in Part L1(c).

Modular buildings

13 Special considerations apply to modular and/ or portable buildings in those situations where the intended life of the building is more than two years.

'Temporary buildings with a planned time of use of two years or less' are exempt from the energy efficiency requirements – see the copy of Regulation 9 on page 7.

14 Where more than 70% of the external envelope is to be created from sub-assemblies manufactured before 06 April 2006 Part L and regulation 17C apply reasonable provision would be to follow the guidance in *Energy Performance Standards for Modular and Portable Buildings*[6].

These sub-assemblies could be obtained from a centrally held stock or the disassembly of buildings on other premises.

Buildings that are exempt from the requirements in Part L

15 The provisions for exempting buildings and building work from the Building Regulations requirements have changed. See the copy of Regulation 9 on page 7.

16 Examples of buildings which are industrial sites and workshops with low energy demand include buildings or parts of buildings designed to be used separately whose purpose is to accommodate industrial activities in spaces where the air is not conditioned. Activities that would be covered include foundries, forging and other hot processes, chemical process, food and drinks packaging, heavy engineering, and storage and warehouses where, in each case, the air in the space is not fully heated or cooled. Whilst not fully heated or cooled these cases may have some local conditioning appliances such as plaque or air heaters or air conditioners to serve people at work stations or refuges dispersed amongst and not separated from the industrial activities.

17 Examples of 'non-residential agricultural buildings with low energy demand' include buildings or parts of buildings that, for instance, are heated for a few days each year to enable plants to germinate but are otherwise unheated.

---

[6]   *Energy Performance Standards for Modular and Portable Buildings,* Modular and Portable Buildings Association (MPBA), 2006. Available from www.mpba.biz

Section 1: Design standards

REGULATIONS

18 Regulations 17A, 17B, 17C and 17E implement Articles 3 and 4 of the Energy Performance of Buildings Directive and state that:

17A. The Secretary of State shall approve a methodology of calculation of the energy performance of buildings.

17B. The Secretary of State shall approve minimum energy performance requirements for new buildings, in the form of target CO_2 emission rates, which shall be based upon the methodology approved pursuant to regulation 17A.

17C. Where a building is erected, it shall not exceed the target CO_2 emission rate for the building that has been approved pursuant to regulation 17B.

17E. In this Part 'building' means the building as a whole or parts of it that have been designed or altered to be used separately.

Target carbon dioxide Emission Rate (*TER*)

19 The Target CO_2 Emission Rate (*TER*) is the minimum energy performance requirement specified in Regulation 17B. It is the mass of CO_2, emitted per year per square metre of the ***total useful floor area*** of the building (kg/m²/year).

20 The *TER* must be calculated using one of the calculation tools included in the methodology for calculating the energy performance of buildings approved by the Secretary of State pursuant to regulation 17A. This approval is given in Annex I in ODPM Circular 04/2006. Those tools include:

a. The Simplified Building Energy Model (SBEM)[7] for those buildings whose design features are capable of being adequately modelled by SBEM; or

b. Other approved software tools. The procedures for approving such software are set out in Annex I of ODPM Circular 04/2006.

As part of the submission to a building control body (see paragraph 25), the applicant must show that the software tool used is appropriate to the application.

21 The *TER* is calculated in two stages as described below.

a. Firstly, use an approved calculation tool to calculate the CO_2 emission rate ($C_{notional}$) from a notional building with specified properties as described in paragraph 22.

b. Secondly, adjust the CO_2 emissions rate calculated in step a) according to the procedure outlined in paragraph 23.

22 The notional building must:

a. be the same size and shape as the actual building; and

b. comply with the energy performance values set out in the detailed definition of the notional building as set out in the SBEM[7] in respect of both the building fabric and the ***fixed building services***. Under the specific circumstances set out in paragraph 76, the ***air permeability*** used in the calculation of the *TER* may be varied from the value set out in the detailed definition of the notional building as set out in the SBEM. Other values must not be varied; and

c. have the same area of vehicle access doors and ***display windows*** as the actual building; and

d. exclude any service that is not a ***fixed building service*** (such as vertical transport systems); and

e. have the same activity areas and classes of building services as in the actual building. The activity areas with their associated classes of building services must be selected from the predefined standard activity areas specified in the SBEM[7]; and

f. be subject to the occupancy times and environmental conditions (temperatures, illuminance, ventilation rate etc.) in each activity area as defined by the standard data associated with the reference schedules; and

It is recognised that in some cases, designers may vary illumination levels in the actual building from that specified in the notional. However, in order to make the comparison on a like for like basis, the notional and actual buildings should deliver the same level of service provision. In this way, the compliance check tests energy efficiency (W/m².100 lux), not energy conservation (W/m²).

g. be subject to the climate defined by the CIBSE Test Reference Year for the site that is most appropriate to the location of the actual building; and

h. assume mains gas as the heating fuel where it is to be used in the actual building, but otherwise assume oil.

i. assume grid mains electricity will be used as the energy source for all other building services.

j. Use the CO_2 emission factors in Table 2.

k. assume the most energy intensive fit-out specifications will be adopted throughout where a building is proposed for approval

---

[7] Simplified Building Energy Model (SBEM) user manual and Calculation Tool, available at www.odpm.gov.uk

excluding *fit-out works* (for example "shell and core" building developments and business park units) and space is to be offered with a range of services options. In addition, any spaces that have the potential for fitting out without air-conditioning should also comply with criterion 3 as if they were not to be air-conditioned.

For example, if a speculative shell and core office building has the potential to be fitted out as heated and naturally ventilated or air conditioned, the BER and TER at completion of the shell and core building works should be based on assuming that air-conditioning will be installed throughout. In addition to this the shell and core building should meet Criterion 3 (limiting solar gain) as if it is not to be air conditioned.

Guidance on ways of showing fit-out works comply is given in ADL2B.

23 The **TER** is obtained from the following formula:

TER = $C_{notional}$ x (1 – improvement factor) x (1 – LZC benchmark), where

a. 'improvement factor' is the improvement in energy efficiency as given in column (a) of Table 1 appropriate to the classes of building services in the actual building. If different areas of the actual building have different classes of building services, then the level of improvement should be calculated by applying the relevant improvement factor to each separate activity area individually.

b. 'LZC benchmark' is the benchmark provision for low and zero carbon (LZC) energy sources as given in column (b) of Table 1 (see paragraph 32 for additional guidance).

This implements the requirement in Article 5 of the Energy Performance of Buildings Directive to give consideration to the incorporation of low and zero carbon energy supply systems before construction starts. Designers can choose to include more renewable systems in their actual building than the LZC benchmark, although the extent to which this extra can be traded off against fabric measures is limited by paragraphs 34 to 39. A lesser renewable systems provision would have to be compensated by enhanced energy efficiency measures.

CRITERION 1 – ACHIEVING AN ACCEPTABLE BUILDING CO_2 EMISSION RATE (BER)

24 To demonstrate that the requirement in Regulation 17C has been met, the actual building's **BER** must be no greater (worse) than the **TER** calculated as set out in paragraphs 19 to 23.

Calculating the CO_2 emissions from the actual building – Regulation 20D

25 The **BER** must be calculated using the same calculation tool as used for establishing the **TER**.

a. The final calculation produced in accordance with Regulation 20D must be based on the building as constructed, incorporating:

  i. any changes to the performance specifications that have been made during construction.

  ii. the measured **air permeability**, ductwork leakage and fan performances as commissioned.

26 Regulation 20D is as follows.

20D.–(1) Subject to paragraph (4), where regulation 17C applies the person carrying out the work shall provide to the local authority a notice which specifies:

a. the target CO_2 emission rate for the building; and

b. the calculated CO_2 emission rate for the building as constructed.

(2) The notice shall be given to the local authority not later than the date on which the notice required by regulation 20B is required to be given.

(3) A local authority is authorised to accept, as evidence that the requirements of regulation 17C would be satisfied if the building were constructed in accordance with an accompanying list of specifications, a certificate to that effect by a person who is registered by:

a. FAERO Limited; or

b. BRE Certification Limited,

in respect of the calculation of CO_2 emission rates of buildings.

Table 1 Improvement factors and LZC benchmarks for use in the TER equation[1]

Building services strategy for the actual building	(a) Improvement factor	(b) LZC benchmark
Heated and naturally ventilated	0.15	0.10
Heated and mechanically ventilated[2]	0.20	0.10
Air conditioned	0.20	0.10

Notes:

1. For example, the **TER** for an air conditioned space would be $C_{notional}$ x (1 – 0.20) x (1 – 0.10) = 0.72 x $C_{notional}$; an improvement over the 2002 standard of 28%.

2. Mechanical ventilation means systems intended to run continuously during occupied hours. This excludes, for instance, intermittent toilet extract fans.

(4) Where such a certificate is given to the local authority:

a. paragraph (1) does not apply; and

b. the person carrying out the work shall provide to the local authority not later than the date on which the notice required by regulation 20B is required to be given a notice which:

    i. states whether the building has been constructed in accordance with the list of specifications which accompanied the certificate; and

    ii. if it has not, lists any changes to the specifications to which the building has been constructed.

27 In addition to this final calculation it would be useful to both builder and building control body if the builder carries out a preliminary calculation before construction starts based on plans and specifications and shares the results. The calculation tool will give a firm indication of whether a design is compliant and it produces a list of those features of the design that are critical to achieving compliance.

BCBs may ask for this information as part of their preparations for checking compliance.

28 In order to determine the **BER**, the CO_2 emission factors[8] in Table 2 should be used.

These figures are now given in terms of CO_2, not carbon to be more in line with the Directive. The entries are therefore different by a factor of 44/12 compared to the ADL2(2002) data.

29 When systems are capable of being fired by more than one fuel, then:

a. For biomass-fired systems rated at greater than 100kW output but where there is an alternative appliance to provide standby, the CO_2 emission factor should be based on the fuel that is normally expected to provide the lead.

This is to encourage biomass systems, but which are often backed up by fossil-fuelled standby plant.

b. For systems rated at less than 100kW output, where the same appliance is capable of burning both biofuel and fossil fuel, the CO_2 emission factor for dual fuel appliances should be used, except where the building is in a smoke control area, when the anthracite figure should be used.

c. In all other cases, the fuel with the highest CO_2 emission factor should be used.

This option is to cover dual fuel systems, where the choice of fuel actually used depends on prevailing market prices.

Table 2 CO_2 emission factors

Fuel	CO_2 emission factor kgCO$_2$/kWh
Natural Gas	0.194
LPG	0.234
Biogas	0.025
Oil	0.265
Coal	0.291
Anthracite	0.317
Smokeless fuel (inc. coke)	0.392
Dual fuel appliances (mineral + wood)	0.187
Biomass	0.025
Grid supplied electricity	0.422
Grid displaced electricity[1]	0.568
Waste heat[2]	0.018

Notes:

1. Grid displaced electricity comprises all electricity generated in or on the building premises by, for instance, PV panels, wind-powered generators, combined heat and power (CHP) etc. The associated CO_2 emissions are deducted from the total CO_2 emissions for the building before determining the **BER**. CO_2 emissions arising from fuels used by the building's power generation system (e.g. to power the CHP engine) must be included in the building CO_2 emissions calculations.

2. This includes waste heat from industrial processes and power stations rated at more than 10MWe and with a power efficiency >35%.

---

[8]   *CO_2 Emission Figures for Policy Analysis*, BRE, July 2005. Available at www.bre.co.uk/filelibrary/CO2EmissionFigures2001.pdf

30 If thermal energy is supplied from a district or community heating or cooling system, emission factors should be determined by considering the particular details of the scheme. Calculations should take account of the annual average performance of the whole system (i.e. the distribution circuits, and all the heat generating plant, including any CHP, and any waste heat recovery or heat dumping). The **BER** submission should be accompanied by a report, signed by a suitably qualified person, detailing how the emission factors have been derived.

Achieving the *TER*

31 Certain management features offer improved energy efficiency in practice. Where these management features are provided in the actual building, the **BER** can be reduced by an amount equal to the product of the factor given in Table 3 and the CO_2 emissions for the system(s) to which the feature is applied.

*For example, if the CO_2 emissions due to electrical energy consumption were 70kgCO$_2$/m^2.year) without power factor correction, the provision of correction equipment to achieve a pf of 0.95 would enable the **BER** to be reduced by 70 x 0.025 = 1.75 kgCO$_2$/(m^2.year).*

32 In appropriate circumstances, LZC energy supply systems such as solar hot water, photovoltaic power, bio-fuels (e.g. wood fuels and oil blends), combined heat and power (at the building or community levels), and heat pumps can make substantial and cost effective contributions to achieving **TER**s. The 'Low or Zero Carbon Energy Sources – Strategic Guide'[9] describes a range of possible systems and how their contribution to the **BER** can be assessed at the feasibility stage.

CRITERION 2: LIMITS ON DESIGN FLEXIBILITY

33 Whilst the approach to complying with Criterion 1 allows considerable design flexibility, Part L requires that reasonable provision should be made to limit heat gains and losses (Part L1(a), and that energy efficient *fixed building services* and effective controls be provided (L1(b)). These requirements would be met by specifying performance standards that are no worse than those given in paragraphs 34 to 62.

*Implementation of these standards alone will NOT achieve the **TER**; better performance will be required in some or all areas to meet the target.*

Design limits for envelope standards

34 This section sets out the design limits for the building fabric to meet requirements L1(a)(i).

U-values

35 U-values shall be determined in accordance with the methods and conventions as set out in BR 443: Conventions for U-value calculations[10]. The CAB/CWCT publication[11] gives guidance on calculating thermal performance factors for curtain walling.

36 Table 4 sets out limits on design flexibility that are considered reasonable for the purposes of achieving the *energy efficiency requirements*:

a. Column (a) sets out limits for area-weighted average U-values for the elements of the stated type.

Table 3 **Enhancement management and control features**

Feature	Adjustment factor
Automatic monitoring and targeting with alarms for out of range values	0.050
Power factor correction to achieve a whole-building power factor of at least 0.90[1]	0.010
Power factor correction to achieve a whole building power factor of at least 0.95[1]	0.025

Notes:

1   The power factor adjustment can only be taken if the whole building power factor is corrected to the level stated. The two levels of power factor correction are alternative values, not additive.

---

9   *Low or Zero Carbon Energy Sources: Strategic Guide*, NBS, 2006.

10   BR 443 *Conventions for U-value Calculations*, BRE, 2006.

11   *The thermal assessment of window assemblies, curtain walling and non-traditional building envelopes.* CWCT, 2006 (ISBN 1 87400 338 6).

The area-weighted average is calculated by summing the UA values of all elements of a given type (e.g. wall elements) and dividing by the total area of those same elements.

b. column (b) gives limits for U-values for individual elements of the stated type.

To minimise condensation risk in localised parts of the envelope. An individual element is defined as those areas of the given element type that have the same construction details. In the case of windows, doors and rooflights, the assessment should be based on the whole unit (i.e. in the case of a window, the combined performance of the glazing and the frame).

37 When comparing against the values in Table 4, the U-value of a window, roof window or rooflight or personnel door can be taken as the value for either:

a. the standard configuration set out in BRE 443; or

b. the particular size and configuration of the actual unit.

SAP 2005[12] Table 6e gives values for different window configurations that can be used in the absence of test data or calculated values.

Table 4 Limiting U-value standards (W/m²·K)

Element	(a) Area-weighted average	(b) For any individual element
Wall	0.35	0.70
Floor	0.25	0.70
Roof	0.25	0.35
Windows[1], roof windows, rooflights[2] and curtain walling	2.2	3.3
Pedestrian doors	2.2	3.0
Vehicle access and similar large doors	1.5	4.0
High usage entrance doors	6.0	6.0
Roof ventilators (inc. smoke vents)	6.0	6.0

Notes:

1   Excluding **display windows** and similar glazing. There is no limit on design flexibility for these exclusions but their impact on CO_2 emissions must be taken into account in calculations.

2   The U-values for roof windows and rooflights in this table are based on the U-value having been assessed with the roof window or rooflight in the vertical position. If a particular unit has been assessed in a plane other than the vertical, the standards given in this Approved Document should be modified by making an adjustment that is dependent on the slope of the unit following the guidance given in BR 443.

[12]   The Government's Standard Assessment Procedure for the energy rating of dwellings, SAP 2005, Defra, 2005.

38 In buildings with high internal gains, a less demanding area weighted average U-value for the glazing may be an appropriate way of reducing overall CO_2 emissions and hence the **BER**. If this case can be made, then the average U-value for windows can be relaxed from the values given in column (a) of Table 4. However values should be no worse than $2.7W/m^2K$. The limit for individual glazing elements given in column (b) should not be exceeded unless there are exceptional circumstances, such as constraints imposed by planning authorities.

Air permeability

39 A reasonable limit for the **design air permeability** is $10m^3/(h.m^2)$ @ 50 Pa. Guidance on some ways of achieving this is given in the TSO publication on robust construction details[13].

*Better standards of **air permeability** are technically desirable in buildings with mechanical ventilation and air conditioning.*

Design limits for building services

40 This section sets out the design limits for **fixed building services** to meet requirement L1(b).

Controls

41 Systems should be provided with appropriate controls to enable the achievement of reasonable standards of energy efficiency in use. In normal circumstances, the following features would be appropriate for heating, ventilation and air conditioning system controls:

a. The systems should be sub-divided into separate control zones to correspond to each area of the building that has a significantly different solar exposure, or pattern, or type of use; and

b. Each separate control zone should be capable of independent timing, and temperature control, and, where appropriate ventilation and air recirculation rate; and

c. The provision of the service should respond to the requirements of the space it serves. If both heating and cooling are provided, they should be controlled so as not to operate simultaneously; and

d. Central plant should only operate as and when the zone systems require it. The default condition should be off.

42 In addition to these general control provisions, the systems should meet specific control and efficiency standards as set out in the paragraphs below.

Energy meters

43 Reasonable provision for energy meters would be to install energy metering systems that enable: at least 90% of the estimated annual energy consumption of each fuel to be assigned to the various end-use categories (heating, lighting etc.). Detailed guidance on how this can be achieved is given in CIBSE TM 39[14]; and

a. the performance of any LZC system to be separately monitored; and

b. in buildings with a **total useful floor area** greater than $1000m^2$, automatic meter reading and data collection facilities.

Heating and hot water service system(s)

44 Reasonable provision for the performance of heating and hot water service systems would be to follow the guidance in the Non-domestic Heating, Cooling and Ventilation Compliance Guide[15], in providing:

a. suitably efficient heating plant; and

b. effective control systems.

The checklists included in the Non-domestic Heating, Cooling and Ventilation Compliance Guide can help in demonstrating that reasonable provision has been made.

Cooling plant

*The carbon emissions associated with the operation of cooling systems are comparatively severe. Reducing solar and internal heat gains and arranging plant and control systems to match the demand effectively over the cooling season can therefore significantly reduce the **BER**.*

45 Reasonable provision for the performance of cooling systems would be to follow the guidance in the Non-domestic Heating, Cooling and Ventilation Compliance Guide in providing:

a. suitably efficient cooling plant; and

b. effective control systems.

The checklists included in the Non-domestic Heating, Cooling and Ventilation Compliance Guide can help in demonstrating that reasonable provision has been made.

Air handling plant

46 Reasonable provision for the performance of air handling plant would be to follow the guidance in the Non-domestic Heating, Cooling and Ventilation Compliance Guide in providing:

a. suitably efficient air handling plant; and

b. effective control systems.

---

[13]   *Limiting Thermal Bridging and Air Leakage: Robust construction details for dwellings and similar buildings*, Amendment 1, TSO, 2002. See www.est.org.uk

[14]   TM 39 *Building Energy Metering*, CIBSE, 2006.

[15]   *Non-domestic Heating, Cooling and Ventilation Compliance Guide*, NBS, 2006.

47 In addition, the system should be capable of achieving a *specific fan power* at 25% of design flow rate no greater than that achieved at 100% design flow rate. Reasonable provision for ventilation system fans rated at more than 1,100 Watts would be to equip them with variable speed drives.

Following this guidance would facilitate commissioning and provide flexibility for future changes of use. The guidance is not applicable to smoke control fans and similar ventilation systems only used in abnormal circumstances.

48 In order to limit air leakage, ventilation ductwork should be made and assembled so as to be reasonably airtight. One way of achieving this would be to comply with the specifications given in HVCA DW/144[16].

Insulation of pipes, ducts and vessels

49 Reasonable provision for compliance with Part L1(a)(ii) would be demonstrated by insulating pipes, ducts and vessels to standards not less than those set out in the Non-domestic Heating, Cooling and Ventilation Compliance Guide.

The TIMSA Guide explains the derivation of the performance standards and how they can be interpreted in practice.

General lighting efficacy in office, industrial and storage areas in all building types

50 For the purposes of this Approved Document, office areas include those spaces that involve predominantly desk-based tasks, including classrooms, seminar rooms and conference rooms, including those in schools.

51 Reasonable provision would be to provide lighting with an average initial efficacy of not less than 45 luminaire-lumens/circuit-Watt as averaged over the whole area of these types of space in the building.

This allows design flexibility to vary the light output ratio of the luminaire and the luminous efficacy of the lamp.

52 The average luminaire-lumens/circuit-Watt is calculated by:

(Lamp lumens x LOR) summed for all luminaires in the relevant areas of the building, divided by the total circuit Watts for all the luminaires where:

Lamp lumens = the sum of the average initial (100 hour) lumen output of all the lamp(s) in the luminaire and

LOR = the light output ratio of the luminaire, i.e. the ratio of the total light output under stated practical conditions to that of the lamp or lamps contained in the luminaire under reference conditions.

*Note that in Approved Document L2B, this equation is modified to include the impact of lighting controls. This is not appropriate in Approved Document L2A, where the calculation tool used to determine the **BER** accounts for the impact of controls.*

General lighting efficacy in all other types of space

53 For lighting systems serving other types of space, it may be appropriate to provide luminaires for which photometric data is not available and/or are lower powered and use less efficient lamps. For such spaces, the requirement would be met if the installed lighting has an average initial (100 hour) lamp plus ballast efficacy of not less than 50 lamp lumens per circuit-Watt.

Controls for general lighting in all types of space

54 Lighting controls should be provided so as to avoid unnecessary lighting during the times when daylight levels are adequate or when spaces are unoccupied.

For safety reasons automatically switched lighting systems should be subjected to risk assessment which may indicate safety should take precedence over energy efficiency.

55 Reasonable provision would be local switches in easily accessible positions within each working area, or at boundaries between working areas and general circulation routes, that are manually operated by the deliberate action of the occupants.

Manual switches include rocker switches, push buttons and pull cords and remote switching devices such as wireless transmitters and telephone handsets.

56 For the purposes of this Approved Document, switches include dimmer switches and switching includes dimming. It would usually be reasonable for dimming to be effected by reducing rather than diverting the energy supply.

57 The distance on plan from any local switch to any luminaire it controls should generally be not more than six metres or twice the height of the luminaire above the floor if this is greater. Where a space is a *daylit space* served by side windows, it would be reasonable for the perimeter row of luminaires to be separately switched.

58 Occupant control of local switching can be supplemented by other controls such as automatic systems which:

a. switch the lighting off when they sense the absence of occupants; or

b. either dim or switch off the lighting when there is sufficient daylight. When installed in appropriate locations, such control systems can make a useful contribution towards reducing the **BER**.

59 A way of meeting the requirement would be to follow the recommendations in BRE Digest 498[18].

[16]   DW/144 *Specification for Sheet Metal Ductwork*, HVCA, 1998.

[17]   *HVAC Guidance for Achieving Compliance with Part L of the Building Regulations*, TIMSA, 2006.

[18]   BRE Digest 498 *Selecting Lighting Controls*, BRE, 2006.

Display lighting in all types of space

60 Reasonable provision for **display lighting** would be to demonstrate that the installed **display lighting** has an average initial (100 hour) efficacy of not less than 15 lamp-lumens per circuit-Watt. In calculating this efficacy, the power consumed by any transformers or ballasts should be taken into account

61 Spaces where **display lighting** is present would normally be expected to also have general lighting used for circulation and for purposes of cleaning and restocking outside public access hours. Paragraphs 50 to 59 apply to this general lighting, depending on the type of space.

Controls for display lighting in all types of space

62 A way of meeting the requirement would be to connect **display lighting** in dedicated circuits that can be switched off at times when people will not be inspecting exhibits or merchandise or attending entertainment events. In a retail store, for example, this could include timers that switch the **display lighting** off outside store opening hours, except for displays designed to be viewed from outside the building through **display windows**.

CRITERION 3: LIMITING THE EFFECTS OF SOLAR GAINS IN SUMMER

Spaces not served by air conditioning systems

For the purposes of this Approved Document, an air conditioning system refers to any system where refrigeration is used to provide cooling for the comfort of occupants.

63 For occupied spaces excluding spaces such as stacks, unoccupied atria intended to drive natural ventilation via buoyancy and spaces adjacent to **display glazing** that are not served by air conditioning systems, provisions should be made to limit solar gains so as to reduce internal temperature rise in summer. This can be done by an appropriate combination of window sizing and orientation, solar protection through shading and other solar control measures, and by using thermal capacity coupled with night ventilation. BR 364[19] and CIBSE AM10[20] offer guidance on strategies to limit solar gain.

*The Building Regulations do not specify minimum daylight requirements although window area has an impact on the **BER**, because energy consumed by electric lighting will increase as window area decreases. On the other hand, conduction heat loss and solar gains in winter would also be reduced. Therefore, when considering the proportions of glazing in the building, the designer should give consideration to the provision of adequate levels of daylight – for guidance on*

daylighting see BS8206 Part 2[21] and NARM technical guidance[22]. Specifying efficient lighting with effective controls will reduce internal gains that will also help to reduce internal temperature rise in summer when daylight availability is at a maximum.

Designers and their clients may wish to go beyond the requirements in the Building Regulations to take more account of the impacts of future global warming on the risks of higher internal temperatures occurring more often. CIBSE TM 36[23] gives guidance on this issue.

64 Reasonable provision would be to show for every occupied space which is not air conditioned that:

a. when the building is subject to the solar irradiances for July as given in the table of design irradiancies in CIBSE Design Guide A[24], the combined solar and internal casual gains (people, lighting and equipment) per unit floor area averaged over the period 0630 to 1630 Solar Time (GMT) is not greater than 35W/m². TM 37[25] gives guidance and supporting data to enable this check to be made and includes an adjustment factor to allow the basic limiting gain of 35W/m² to be adjusted, dependent upon the location of the building; or

This is Table 2.30 in the 2006 edition of the CIBSE Design Guide. Note that the table is expressed in solar time i.e. GMT. When calculating the gains for a perimeter area with side windows, it would be normal to calculate the gains over the area within 6m from the window wall. This simplified guidance would enable compliance with the Building Regulations requirements but designers may wish to satisfy themselves that the proposed design also meets clients' expectations.

b. the operative temperature (the temperature index for thermal comfort as used in CIBSE Guide A) in the conditioned space does not exceed a threshold for more than a reasonable number of occupied hours per year when the building is tested against the CIBSE Design Summer Year appropriate to the building location; or

---

[19] BR 364 *Solar Shading of Buildings*, BRE, 1999 (reprinted 2001).

[20] AM10 *Natural ventilation in non-domestic buildings*, CIBSE, 2005.

[21] BS 8206:1992 *Lighting for Buildings, Code of Practice for Daylighting*.

[22] *Use of rooflights to satisfy the 2002 Building Regulations for the Conservation of Fuel and Power*, NARM.

[23] TM 36 *Climate change and the indoor environment: impacts and adaptation*, CIBSE, 2006.

[24] CIBSE Guide A: *Environmental Design*. CIBSE, 2006.

[25] TM 37 *Design for Improved Solar Shading Control*, CIBSE, 2006.

The threshold temperature and the reasonable number of hours will depend on the activities within the space. In order to meet Workplace Regulations, clients, designers and health and safety inspectors should agree appropriate values that would be considered tolerable by occupiers. For example, CIBSE suggest that for office-type spaces, the number of occupied hours with dry resultant temperatures over 28ºC should not exceed 1% of the annual occupied period.

c. For school buildings, Building Bulletin 101[26] specifies overheating criteria and provides guidance on methods to demonstrate that reasonable provision has been made to control excessive solar gains.

Spaces served by air conditioning systems

65 For spaces served by air conditioning systems, reasonable provision for the control of excessive solar gains is demonstrated by meeting the **TER**. However, if solar gains are controlled to the limits as determined by paragraph 63a), cooling energy demand will be moderate, and it will be easier to achieve the **TER**.

*The **TER** is based on a notional building with modest amounts of glazing. Building designers that prefer more glazing that allows greater solar gain will have to compensate through enhanced energy efficiency measures in other aspects of the design.*

[26] *Ventilation of School Buildings*, Building Bulletin 101, School Building and Design Unit. Department of Education and Skills, 2006. www.teachernet.gov.uk/iaq

SECTION 2: Quality of construction

CRITERION 4 – QUALITY OF CONSTRUCTION AND COMMISSIONING

66 Buildings should be constructed and equipped so that performance is consistent with the predicted **BER**. As indicated in paragraph 26a), a final calculation of the **BER** is required to reflect:

a. any changes in performance between design and construction; and

b. the achieved **air permeability**, ductwork leakage and commissioned fan performance.

*The report referred to in paragraph 27. would assist **BCBs** in checking the key features of the design are included as specified during the construction process.*

Building fabric

67 The building fabric should be constructed to a reasonable quality so that:

a. The insulation is reasonably continuous over the whole building envelope; and

b. the **air permeability** is within reasonable limits.

Continuity of insulation

68 The building fabric should be constructed so that there are no reasonably avoidable thermal bridges in the insulation layers caused by gaps within the various elements, at the joints between elements and at the edges of elements such as those around window and door openings.

69 Reasonable provision would be to:

a. Adopt design details such as

   i. For construction styles similar to **dwellings**, details from *Limiting thermal bridging and air leakage*; or

These might apply to small-scale buildings such as shops and community centres etc.

   ii. For cladding systems, to adopt the guidance given in the MCRMA Technical Note[27]; or

b. to demonstrate that the specified details deliver an equivalent level of performance using the guidance in BRE IP 1/06[28].

70 In addition, the builder should have an appropriate system of site inspection in place to give confidence that the construction procedures achieve the required standards of consistency. For those using the accredited details approach (paragraph 69a)) a way of achieving this would be to produce a report demonstrating that the construction checklists such as those included in the accredited design details publication have been completed and show satisfactory results.

*It would help builders and **BCBs** if such reports are signed by a suitably qualified person.*

Air permeability and pressure testing

71 Regulation 20B states that:

20B.–(1) This regulation applies to the erection of a building in relation to which paragraph L1(a)(i) of Schedule 1 imposes a requirement.

(2) Where this regulation applies, the person carrying out the work shall, for the purpose of ensuring compliance with regulation 17C and paragraph L1(a)(i) of Schedule 1:

a. ensure that:

   i. pressure testing is carried out in such circumstances as are approved by the Secretary of State; and

   ii. the testing is carried out in accordance with a procedure approved by the Secretary of State; and

b. subject to paragraph (5), give notice of the results of the testing to the local authority.

(3) The notice referred to in paragraph (2)(b) shall:

a. record the results and the data upon which they are based in a manner approved by the Secretary of State; and

b. be given to the local authority not later than seven days after the final test is carried out.

(4) A local authority is authorised to accept, as evidence that the requirements of paragraph (2)(a)(ii) have been satisfied, a certificate to that effect by a person who is registered by the British Institute of Non-destructive Testing in respect of pressure testing for the air tightness of buildings.

(5) Where such a certificate contains the information required by paragraph (3)(a), paragraph (2)(b) does not apply.

72 The approved circumstances under which the Secretary of State requires pressure testing to be carried out are set out in paragraph 74.

73 The approved procedure for pressure testing is given in the ATTMA publication 'Measuring Air Permeability of Building Envelopes'[29]. The manner approved for recording the results and the data on which they are based is given in section 4 of that document.

---

[27] *Guidance for design of metal cladding and roofing to comply with Approved Document L2*, MCRMA. www.mcrma.co.uk

[28] IP 1/06 *Assessing the Effects of Thermal Bridging at Junctions and Around Openings in the external elements of buildings*, BRE, 2006.

[29] *Measuring Air Permeability of Building Envelopes*, Air Tightness Testing and Measurement Association (ATTMA), 2006.

74 All buildings that are not **dwellings** (including extensions which are being treated as new buildings for the purposes of complying with Part L) must be subject to pressure testing, with the following exceptions:

a. buildings less than 500m² **total useful floor area**; in this case the developer may choose to avoid the need for a pressure test provided that the **air permeability** used in the calculation of the **BER** is taken as 15m³/hour.m²) at 50Pa.

Compensating improvements in other elements of the building fabric and building services will be needed to keep the BER no worse than the TER.

b. factory-made modular buildings where no site assembly work is needed; provided that the particular module type has been subjected to an in-situ test programme and certified by an approved pressure testing firm as having satisfactory **design air permeability** and that this is routinely achieved on site.

Site based testing is necessary to demonstrate the building is sufficiently robust to resist flexure during lifting and transportation.

c. Large extensions (whose compliance with Part L is being assessed as if they were new buildings, Approved Document L2B refers) where sealing off the extension from the existing building is impractical. The ATTMA publication gives guidance both on how extensions can be tested and situations where pressure tests are inappropriate. Where it is agreed with the **BCB** that testing is impractical, the extension should be treated as a large, complex building, with the guidance in paragraph 74d) then applying.

d. Large complex buildings, where due to building size or complexity, it may be impractical to carry out pressure testing of the whole building. The ATTMA publication indicates those situations where such considerations might apply. Before adopting this approach developers must produce in advance of construction work in accordance with the approved procedure a detailed justification of why pressure testing is impractical. This should be endorsed by a suitably qualified person such as a Competent Person approved for pressure testing. In such cases, a way of showing compliance would be to appoint a suitably qualified person to undertake a detailed programme of design development, component testing and site supervision to give confidence that a continuous air barrier will be achieved. In such cases it would not be reasonable to claim air **permeability** better than 5.0m³/hour.m² at 50Pa has been achieved.

e. Compartmentalised buildings; where buildings are compartmentalised into self-contained units with no internal connections it may be impractical to carry out whole building pressure tests. In such cases reasonable provision would be to carry out a pressure test on a

representative area of the building as detailed in the ATTMA guidance. In the event of a test failure, the provisions of paragraphs 75 and 76 would apply, but it would be reasonable to carry out a further test on another representative area to confirm that the expected standard is achieved in all parts of the building.

One example of a suitably qualified person would be an ATTMA member. The 5.0m³/hour.m² at 50Pa limit has been set because at higher standards, the actual level of performance becomes too vulnerable to single point defects in the air barrier.

75 Compliance with the requirement in Part L1(a)(i) would be demonstrated if:

a. the measured **air permeability** is not worse than the limit value set out in paragraph 39; and

b. the **BER** calculated using the measured **air permeability** is not worse than the **TER**.

*If it proves impractical to meet the design **air permeability**, any shortfall must be compensated through improvements to subsequent fit-out activities. Builders may therefore wish to schedule pressure tests early enough to facilitate remedial work on the building fabric, e.g. before false ceilings are up.*

Consequences of failing a pressure test

76 If satisfactory performance is not achieved, then remedial measures should be carried out on the building and new tests carried out until the building achieves the criteria set out in paragraph 75. For buildings with a **total useful floor area** of less than 1000m² and in the period up to 31 October 2007, if the initial test result is unsatisfactory, reasonable provision would be to:

a. carry out remedial measures such that on retest, a result was achieved that showed:

   i. an improvement of 75% of the difference between the initial test result and the **design air permeability**; or

   ii. if less demanding, a test result within 15% of the **design air permeability**; and

*This period of easement is given in recognition that it will take time for builders to master the techniques for more airtight construction. As an example, if an initial test result was 18.0, and the **air permeability** standard was 8.0, the modified pass level in the period to October 2007 for subsequent tests following remedial action would be [18.0 − 0.75 x (18.0 − 8.0)] = 10.5m³/hour/m² at 50 Pa. However, if the initial test result was 9.5, the modified pass level in the period to October 2007 for subsequent tests following remedial action would be 8.0 x 1.15 = 9.2.*

b. revise the **TER** by substituting the measured **air permeability** from paragraph 76a) for the value given in the detailed definition of the notional building as set out in the SBEM and demonstrate that the **BER** is no worse than the revised **TER**.

COMMISSIONING OF THE BUILDING SERVICES SYSTEMS

77 The building services systems should be commissioned so that at completion, the system(s) and their controls are left in working order and can operate efficiently for the purposes of the conservation of fuel and power. Regulation 20C states that:

20C.–(1) This regulation applies to building work in relation to which paragraph L1(b) of Schedule 1 imposes a requirement, but does not apply where the work consists only of work described in Schedule 2B.

(2) Where this regulation applies the person carrying out the work shall, for the purpose of ensuring compliance with paragraph L1(b) of Schedule 1, give to the local authority a notice confirming that the fixed building services have been commissioned in accordance with a procedure approved by the Secretary of State.

(3) The notice shall be given to the local authority:

a. not later than the date on which the notice required by regulation 15(4) is required to be given; or

b. where that regulation does not apply, not more than 30 days after completion of the work.

78 The procedure approved by the Secretary of State is that set out in:

a. CIBSE Commissioning Code M on Commissioning Management[30]; and

This provides guidance on the overall process and includes a schedule of all the relevant guidance documents relating to the commissioning of specific building services systems.

b. The procedures for leakage testing of ductwork are given in paragraph 80.

79 The notice should include a declaration confirming that:

a. a commissioning plan has been followed so that every system has been inspected and commissioned in an appropriate sequence and to a reasonable standard; and

b. the results of tests confirm that the performance is reasonably in accordance with the actual building designs, including written commentaries where excursions are proposed to be accepted.

It would be helpful to BCBs if such declarations were to be signed by someone suitably qualified by relevant training and experience. A way of achieving this would be to employ a member of the Commissioning Specialists Association or the Commissioning Group of the HVCA in respect of HVAC systems or a member of the Lighting Industry Commissioning Scheme in repect of fixed internal or external lighting.

Air leakage testing of ductwork

80 Ductwork leakage testing should be carried out in accordance with the procedures set out in HVCA DW/143[31] on systems served by fans with a design flow rate greater than 1m³/s and for those sections of ductwork where:

a. the pressure class is such that DW/143 recommends testing; and

b. the **BER** calculation assumes a leakage rate for a given section of ductwork that is lower than the standard defined in DW/144 for its particular pressure class. In such cases, any low pressure ductwork should be tested using the DW/143 testing provisions for medium pressure ductwork.

Membership of the HVCA specialist ductwork group or the Association of Ductwork Contractors and Allied Services could be a way of demonstrating suitable qualifications for this testing work.

81 If a ductwork system fails to meet the leakage standard, remedial work should be carried out as necessary to achieve satisfactory performance in re-tests and further ductwork sections should be tested as set out in DW/143.

---

[30] CIBSE Code M: *Commissioning Management*, CIBSE, 2003, ISBN 1 90328 733 2.

[31] DW/143 *A Practical Guide to Ductwork Leakage Testing*, HVCA, 2000.

Section 3: Operating and maintenance instructions

CRITERION 5 – PROVIDING INFORMATION

82 In accordance with Requirement L1(c), the owner of the building should be provided with sufficient information about the building, the *fixed building services* and their maintenance requirements so that the building can be operated in such a manner as to use no more fuel and power than is reasonable in the circumstances.

Building log-book

83 A way of showing compliance would be to produce information following the guidance in CIBSE TM 31 Building Log Book Toolkit[32]. The information should be presented in templates as or similar to those in the TM. The information could draw on or refer to information available as part of other documentation, such as the Operation and Maintenance Manuals and the Health and Safety file required by the CDM Regulations.

84 The data used to calculate the *TER* and the *BER* should be included in the log-book.

It would also be sensible to retain an electronic copy of the input file for the energy calculation to facilitate any future analysis that may be required by the owner when altering or improving the building.

---

[32]   TM 31 *Building Log Book Toolkit*, CIBSE, 2006.

Section 4: Model designs

85 Some builders may prefer to adopt model design packages rather than to engage in design for themselves. Such model packages of fabric U-values, boiler seasonal efficiencies, window opening allowances etc. would achieve compliant overall performance within certain constraints. The construction industry may develop model designs for this purpose and make them available on the Internet at www.modeldesigns.info

86 It will still be necessary to demonstrate compliance in the particular case by going through the procedures described in paragraphs 7 to 12.

Section 5: Definitions

87 For the purposes of this Approved Document, the following definitions apply.

88 *Air permeability* is the physical property used to measure airtightness of the building fabric. It is defined as air leakage rate per envelope area at the test reference pressure differential across the building envelope of 50 Pascal (50N/m²). The envelope area of the building, or measured part of the building, is the total area of all floors, walls and ceilings bordering the internal volume subject to the test. This includes walls and floors below external ground level. Overall internal dimensions are used to calculate this area and no subtractions are made for the area of the junctions of internal walls, floors and ceilings with exterior walls, floors and ceilings.

89 *BCB* means Building Control Body: a local authority or an approved inspector.

90 *BER* is the Building CO_2 Emission Rate.

91 *Daylit space* means any space:

a. within 6m of a window wall, provided that the glazing area is at least 20% of the internal area of the window wall; or

b. below rooflights and similar provided that the glazing area is at least 10% of the floor area. The normal light transmittance of the glazing should be at least 70%, or, if the light transmittance is reduced below 70%, the glazing area could be increased proportionally.

92 *Design air permeability* is the value of *air permeability* selected by the building designer for use in the calculation of the *BER*.

93 *Display window* means an area of glazing, including glazed doors, intended for the display of products or services on offer within the building, positioned:

a. at the external perimeter of the building;

b. at an access level and immediately adjacent to a pedestrian thoroughfare.

There should be no permanent workspace within one glazing height of the perimeter. Glazing more than 3m above such an access level should not be considered part of a *display window* except:

a. Where the products on display require a greater height of glazing;

b. In existing buildings, when replacing *display windows* that already extend to a greater height;

c. In cases of building work involving changes to the façade and glazing requiring planning consent, where planners should have discretion to require a greater height of glazing, e.g. to fit in with surrounding buildings or to match the character of the existing façade.

It is expected that *display windows* will be found in buildings in use classes A1, A2, A3 and D2 as detailed in Table 6.

Table 6 **Building classes**

Class	Use
A1	Shops including retail-warehouse, undertakers, showrooms, post offices, hairdressers, shops for sale of cold food for consumption off premises
A2	Financial and professional services banks, building societies, estate and employment agencies, betting offices
A3	Food and drink restaurants, pubs, wine bars, shops for sale of hot food for consumption off premises
D2	Assembly and leisure cinemas, concert halls, bingo halls, casinos, sports and leisure uses

94 *Display lighting* means lighting intended to highlight displays of exhibits or merchandise, or lighting used in spaces for public leisure and entertainment such as dance halls, auditoria, conference halls, restaurants and cinemas.

95 *Dwelling* means a self-contained unit designed to accommodate a single household.

Rooms for residential purposes *are not* ***dwellings*** *so Approved Document L2A is applicable to their construction.*

96 *Emergency escape lighting* means that part of emergency lighting that provides illumination for the safety of people leaving an area or attempting to terminate a dangerous process before leaving an area.

97 *Energy efficiency requirements* means the requirements of Regulations 4A, 17C and 17D and Part L of Schedule 1.

98 *Fit-out work* means that work needed to complete the partitioning and building services within the external fabric of the building (the shell) to meet the specific needs of incoming occupiers. ***Fit-out work*** can be carried out in whole or in parts:

a. In the same project and time frame as the construction of the building shell; OR

b. At some time after the shell has been completed.

99 *Fixed building services* means any part of, or any controls associated with,

a. fixed internal or external lighting systems but does not include emergency escape lighting or specialist process lighting; or

b. fixed systems for heating, hot water service, air conditioning or mechanical ventilation.

100 *High usage entrance door* means a door to an entrance primarily for the use of people that is expected to experience large traffic volumes, and where robustness and/or powered operation is the primary performance requirement. To qualify as a ***high usage entrance door***, the door should be equipped with automatic closers, and except where operational requirements preclude, be protected by a lobby.

101 *Room for residential purposes* is defined in Regulation 2 (1) as follows:

Room for residential purposes means a room, or a suite of rooms, which is not a dwelling-house or a flat and which is used by one or more persons to live and sleep and includes a room in a hostel, an hotel, a boarding house, a hall of residence or a residential home, whether or not the room is separated from or arranged in a cluster group with other rooms, but does not include a room in a hospital, or other similar establishment, used for patient accommodation and, for the purposes of this definition, a 'cluster' is a group of rooms for residential purposes which is:

a. separated from the rest of the building in which it is situated by a door which is designed to be locked; and

b. not designed to be occupied by a single household.

102 *Specialist process lighting* means lighting intended to illuminate specialist tasks within a space, rather than the space itself. It could include theatre spotlights, projection equipment, lighting in TV and photographic studios, medical lighting in operating theatres and doctors' and dentists' surgeries, illuminated signs, coloured or stroboscopic lighting, and art objects with integral lighting such as sculptures, decorative fountains and chandeliers.

103 *TER* is the Target CO_2 Emission Rate.

104 *Total useful floor area* is the total area of all enclosed spaces measured to the internal face of the external walls, that is to say it is the gross floor area as measured in accordance with the guidance issued to surveyors by the RICS. In this convention:

a. the area of sloping surfaces such as staircases, galleries, raked auditoria, and tiered terraces should be taken as their area on plan; and

b. areas that are not enclosed such as open floors, covered ways and balconies are excluded.

Appendix A: Compliance checklist

1 The following table provides a checklist of the evidence that could be compiled to facilitate for developers and BCBs the processes of demonstrating compliance with Part L. The checklist prompts for the evidence that needs to be provided to allow the check to be made and who could produce the evidence.

2 For most steps, the evidence could be provided by an approved Competent Person or a suitably qualified person acting for the developer and may be accepted at face value at the discretion of the BCB dependent upon the credentials of the person making the declaration. In the checklist, the 'Produced by' column indicates the expected source of the evidence, and the header and footer blocks of the checklist allow opportunity for the credentials of those submitting the evidence to be declared. If the evidence is not produced by an approved Competent Person, BCBs have the discretion to accept evidence from other groups of appropriately qualified and/or experienced individuals.

3 The final two columns enable recording the results of the checks.

4 As an aid to monitoring during construction and compliance checking two versions of the checklist could be produced, one for the building as designed, the other for the building as constructed (see paragraph 26). The parts of the checklist that are not relevant are shown on the checklist as N/A.

5 Editable versions of the checklist can be downloaded from the ODPM website.

6 Other than the CO_2 target that is mandatory, the other checks represent reasonable provision in normal circumstances. In unusual circumstances, alternative limits may represent reasonable provision, but this would have to be demonstrated in the particular case.

Checklist for a new building that is not a dwelling

Site reference		Plot reference		
Developer		Contact		☎
Building Control Body		Contact		☎
Calculation outputs produced by:		Contact		☎
Evidence of competency *(e.g. Part L competent Person, Authorised SAP Assessor)*				

No.	Check	Evidence	Produced by	Design OK?	As built OK?
1	**Criterion 1: predicted carbon dioxide emission from proposed dwelling does not exceed the target**				
1.1	Calculated CO_2 emission rate from notional building $kgCO_2/m^2$. annum	Standard output from approved software	Approved Competent Person or Developer*	N/A	N/A
1.2	Improvement factor	From Table 1	Developer*		
1.3	LZC benchmark	From Table 1	Developer*		
1.4	**TER** $(kgCO_2/m^2.a)$	Standard output from approved software	Developer*		
1.5	**BER** $(kgCO_2/m^2.a)$	Standard output from approved software	Approved Competent Person or Developer*		
1.6	Are emissions from building less than or equal to the target?	Compare TER and BER	Approved Competent Person or Developer*		
1.7	Are as built details the same as used in **BER** calculations?	Declaration	Developer*	N/A	
2	**Criterion 2: the performance of the building fabric and the building services systems should be no worse than the limits on design flexibility**				
2a	**Building fabric**				
2.1	Are U-values better than the limits on design flexibility?	Schedule of U-values produced as standard output from accredited software	Approved Competent Person or Developer*		
2.2	Is **air permeability** no greater than the worst acceptable standard?	Standard output from approved software	Approved Competent Person or Developer*		
2b	**Fixed building services**				
2.3	Are all building services standards acceptable?	Schedule of system efficiencies produced as standard output from Approved software	Approved Competent Person or Developer*		
2.4	Does fixed internal lighting comply with paragraphs to 49 to 61?	Schedule of installed fixed internal lighting	Developer*		

2.5	Energy meters installed in accordance with GIL 65?	Metering strategy document	Developer*		
3	**Criterion 3: spaces without air conditioning have appropriate passive control measures to limit the effects of solar gain**				
3.1	Method of showing compliance: para64a, $G \leq 35W/m^2$, or para 64b, operative temperature $t > 28^oC$ for $\leq X$ hours per year, or para 64c, BB101?	Schedule for each zone	Developer*		
4	**Criterion 4: the performance of the building, as built, is consistent with the BER**				
4a	**Building fabric**				
4.1	Have the key features of the design been included (or bettered) in practice?	List of key features produced by approved software to facilitate sample checking by BCB	Building control body	N/A	
4.2	Is the level of thermal bridging acceptable?	Schedule of accredited details used and their reference codes; or Evidence that details adopted deliver equivalent performance	Developer*		
4.3	Has satisfactory documentary evidence of site inspection checks been produced?	Completed proformas showing checklists have been completed	Developer*		
4.4	***Design air permeability*** $(m^3/(h.m^2)$ at 50Pa)	Standard output from approved software	Approved Competent Person or Developer		N/A
4.4	***Design air permeability*** $(m^3/(h.m^2)$ at 50Pa)	Standard output from approved software	Approved Competent Person or Developer		N/A
4b	**Commissioning of the fixed building services**				
4.5	Has evidence been provided that demonstrates that the ***design air permeability*** has been achieved satisfactorily?	Pressure test results in comparison to design svalue; or Report on agreed programme of design development and component testing; or Modular buildings type test results	Developer*	N/A	
4.6	Has commissioning been completed satisfactorily?	Commissioning report submitted in accordance with CIBSE Code M?	Developer*	N/A	

4.6	Has commissioning been completed satisfactorily?	Commissioning report submitted in accordance with CIBSE Code M?	Developer*	N/A	
4.7	Has evidence been provided that demonstrates that the ductwork is sufficiently airtight?	Report confirming that the results of the leakage tests are in line with the leakage specification	Developer*	N/A	
5	**Criterion 5: providing information**				
5.1	Has a suitable building log-book been prepared?	Completed CIBSE TM 31 template (or equivalent)	Developer*	N/A	

Schedule of supporting competencies

No.	* Organisation producing evidence	☎		Evidence of competency
2.4				
4.5				
4.6				
4.7				

Documents referred to

Air Tightness Testing and Measurement Association (ATTMA)
www.attma.org

Measuring air permeability of building envelopes, 2006.

BRE
www.bre.co.uk

BR 262 *Thermal insulation: avoiding risks,* 2001. ISBN 1 86081 515 4

BR 364 *Solar shading of buildings*, 1999 (reprinted 2001).

BR 443 *Conventions for U-value calculations*, 2006. (Downloadable from www.bre.co.uk/uvalues.)

BRE Digest 498 *Selecting lighting controls*, 2006. ISBN 1 86081 905 2

Information Paper IP1/06 *Assessing the effects of thermal bridging at junctions and around openings in the external elements of buildings*, 2006. ISBN 1 86081 904 4

Delivered energy emission factors for 2003. (Available at www.bre.co.uk/filelibrary/2003emissionfactorupdate.pdf.)

CO_2 emission figures for policy analysis, July 2005. (Available at www.bre.co.uk/filelibrary/co2emissionfigures2001.pdf.)

Simplified Building Energy Model (SBEM) user manual and Calculation Tool. (Available at www.odpm.gov.uk.)

Centre for Window and Cladding Technology
www.cwct.co.uk

Thermal assessment of window assemblies, curtain walling and non-traditional building envelopes, 2006. ISBN 1 87400 338 6

CIBSE
www.cibse.org

CIBSE Commissioning Code M *Commissioning Management*, 2003. ISBN 1 90328 733 2

CIBSE Guide A *Environmental Design*, 2006. ISBN 1 90328 766 8

AM 10 *Natural ventilation in non-domestic buildings*, 2005. ISBN 1 90328 756 1

TM 31 *Building Log Book Toolkit*, CIBSE, 2006. ISBN 1 90328 771 5

TM 36 *Climate change and the indoor environment: impacts and adaptation*, 2005. ISBN 1 90328 750 2

TM 37 *Design for improved solar shading control*, 2006. ISBN 1 90328 757 X

TM 39 *Building energy metering*, 2006. ISBN 1 90328 707 7

Department of the Environment, Food and Rural Affairs (Defra)
www.defra.gov.uk

The Government's Standard Assessment Procedure for energy rating of dwellings, SAP 2005.
(Download from www.bre.co.uk/sap2005.)

Department of Transport, Local Government and the Regions (DTLR)

Limiting thermal bridging and air leakage: Robust construction details for dwellings and similar buildings, Amendment 1. Published by TSO, 2002. ISBN 0 11753 631 8
(Download from Energy Saving Trust (EST) website on http://portal.est.org.uk/housingbuildings/calculators/robustdetails/.)

Department for Education and Skills (DfES)

Building Bulletin 101 *Ventilation of School Buildings,* School Building and Design Unit, 2005. (Download from www.teachernet.gov.uk/iaq.)

Health and Safety Executive (HSE)
www.hse.gov.uk

L24 *Workplace Health, Safety and Welfare: Workplace (Health, Safety and Welfare) Regulations 1992, Approved Code of Practice and Guidance, The Health and Safety Commission*, 1992. ISBN 0 71760 413 6

Heating and Ventilating Contractors Association

DW/143 *A practical guide to ductwork leakage testing*, 2000. ISBN 0 90378 330 4

DW/144 *Specification for sheet metal ductwork*, 1998. ISBN 0 90378 327 4

Metal Cladding and Roofing Manufacturers Association
www.mcrma.co.uk

Guidance for design of metal cladding and roofing to comply with Approved Document L2.

Modular and Portable Buildings Association (MPBA)
www.mpba.biz

Energy performance standards for modular and portable buildings, 2006.

National Association of Rooflight Manufacturers
www.narm.org.uk

Use of rooflights to satisfy the 2002 Building Regulations for the Conservation of Fuel and Power.

NBS (on behalf of ODPM)
www.thebuildingregs.com

Non-domestic Heating, Cooling and Ventilation Compliance Guide, 2006. ISBN 1 85946 226 X

Low or Zero Carbon Energy Sources: Strategic Guide, 2006. ISBN 1 85946 224 3

Thermal Insulation Manufacturers and Suppliers Association (TIMSA)
www.timsa.org.uk

HVAC Guidance for Achieving Compliance with Part L of the Building Regulations, 2006.

Legislation

SI 1991/1620 Construction Products Regulations 1991.

SI 1992/2372 Electromagnetic Compatibility Regulations 1992.

SI 1994/3051 Construction Products (Amendment) Regulations 1994.

SI 1994/3080 Electromagnetic Compatibility (Amendment) Regulations 1994.

SI 1994/3260 Electrical Equipment (Safety) Regulations 1994.

SI 2001/3335 Building (Amendment) Regulations 2001.

SI 2006/652 Building And Approved Inspectors (Amendment) Regulations 2006.

Standards referred to

BS 8206-2:1992 Lighting for buildings. Code of
practice for daylighting.